TSUNAMI FROM SOLARIS

ESSAYS ON POETRY

Aase Berg

Translated by Johannes Göransson and Joyelle McSweeney

Action Books, Notre Dame, IN
2019

SALVO SERIES #4

ACTION BOOKS

Joyelle McSweeney and Johannes Göransson, Editors
Natasha Ali, Maxime Berclaz and Sebastian Bostwick, 2018-2020 Editorial Assistants
Anne Malin Ringwalt, 2017-2019 Editorial Assistant
Andrew Shuta, Book Design

Tsunami From Solaris, Essays on Poetry

ISBN 9780900575839

Library of Congress Control Number: 2019933240

Action Books is housed in the Creative Writing Program, Department of English, College of Arts and Letters, University of Notre Dame.

Send us mail at 233 Decio Hall, Notre Dame, IN, 46556 or visit us on-line at actionbooks.org.

Table of Contents

THE WOUNDED PATH TOWARD AN INTERESTING LIFE

Scars lie waiting for new attacks. They are not at all as safely and reliably passive as one might expect. Some scars don't mark an end but rather a new beginning for the wound's history. Those scars are not so much a sign of something that has already happened – i.e., "the catastrophe of our time!" – as entrances toward new, open, pink-glowing flesh wounds. Small children often find an aggressive joy in picking away at scabs. One can go even further: like HP Lovecraft, one can dig down into the catacombs and underground muck to find in the earth's depths new, dream-provoking lifeforms. Certain poets are constantly repeating this movement of digging into the wounds, plundering the graves. Eva Kristina Olsson is one such poet. The goal of such degenerate activity is perhaps to pursue "the excluded middle" or simply a somewhat interesting life.

Chronological Healing

Let's begin with some conventional thinking. Scars are wounds that clot and then heal. A scar is also a sign indicating something has happened. Often associated with experience and respect, a scar carries the connotation of "war-wounds" or "hard-earned wisdom." This implies that something *has been learned* from a past catastrophe. The wound itself implies survival, since pain indicates a living being. But on a biological level, a scar also carries within it the potential for destruction:

> Scar formation in an inner organ can gradually lead to symptoms of illness as the scar tissue shrinks and contracts. Scar formation around the urinary tract—for example, after gonorrhea—can cause a fusing of the tract, which will lead to difficulty urinating. Scar formation in the esophagus can erode the mucous membrane;

> it can even cause the esophagus to fuse, which makes it hard to swallow food.
>
> —*Medical Encyclopedia of the Swedish Red Cross, 1941*

Scars can fuse and close holes, including both "negative" holes (such as wounds) and "positive" holes (such as the holes needed for nutritional intake). One can consider fossils an extreme form of scar, as well as mold: mold is what happens to food dishes that try to heal and "grow shut" over a long period of time. Even shadows can be viewed as a kind of scar, a kind of fleeting impression or memory of the flesh's career as a foreign object in the surrounding material. Further, the charred land left behind after forest fires makes a kind of scar in the terrain. Those areas where new species of trees are planted instead of the kind that has been chopped down or burnt conjures the artificial scarring of skin transplants.

This is how the land constantly stuffs its own wounds in a stubborn, bullheaded healing process. It's the hamster wheel of the life force, only temporarily interrupted by life-threatening injuries and devastations. What's the point, one may ask oneself. Or: Are there any possible alternative entrances or exits in these strange fusings, these struggling states?

In difference to the wound, the scar has a temporal dimension. When the wound is open and when the poem by certain poets takes shape, there is no time, just a forever screaming now. The words, to the extent that they have articulable meaning, can keep the inside inside and the outside outside. But the sum of the words does not make a poem – there is also an invisible putty, a spirit. The poem in and beyond the words is a gaping hole, a chasm, or void: a wound. In the poetic now, experience is more important than documentation, summarization or delineation. The poem is formless, a protoplasm, a jellyfish, an amoeba that glides around and ignores the human race's grave chronologies. At the same time as the poem is created, it tears up the now in pain, in *nowness*. The poem is in many cases a bubbling becoming-wound. No healing process can take place while it drips from the pen.

In cases when the now, despite everything, heals, a chronology takes shape: the scar stabilizes the individual memory as well as the collective history. This is a sign that the sense-derangement of the wound is behind us and that time has passed. The scar becomes a chronicle, a structure, a small box of hardened skin where one can maintain one's traumatic memories. One can put a label on the scar: "got into a propeller," "knife fight," "guinea pig bite." There is a lovely safety in this form of passive chronology. And there's a necessity in this logic that can't be dismissed: Who can live, in the flesh, a life that is one eternal, open, screaming wound?

But what comes after the scar? Mold? Fossilization, petrification? The fusing, the strangling? The body's death in the coffin, deep down in Lovecraft's wonderful underground soil? Yes, death, as usual. At the end of this process, one stands around and grows depressed. Doesn't it get more interesting than this?

To Break into the Wound

One exciting example of scar-breakage can be found in Eva Kristina Olsson's poetry: here the scar is torn off and becomes a wound again. Here's how it sounds in the book *Song Edges*:

the edges flee
i am not in the place
around the throat hangs a stone
the edges flee
i am in place
openly shaking

The edges here function as scar formations, signs, structure. The stone is pulled down and "the edges flee." Down in the hole, "i am in the place," down in the wound, the now. The "place" no longer has any edges: the outside runs in, the inside runs out – the boundary that made "the place" into a place and the wound into a wound has left the field open to *all eternity*. The wound has been scratched open here (except in the micro-state, in the words/scars themselves), and the boundaries are about to loosen and fall off.

But in her poetry, Olsson is trying to get a handle on the edges, to the extent that the poem is still situated in relationship to them. It's just a matter of time before one gives up hope and finally lets go of the demand for an edge, to instead give oneself totally to the wound. This is where one finds oneself truly in the site of "openly shaking." The place gives way. In Olsson's most recent collection, *The Sacrifice*, cultural and cosmic outlines have been thrown into an irreversible state of dissolution. We find ourselves in the centrifugal flesh, in a chaos of pre-oedipal nowness, in the absolutely naked wound's rabid intestinal mess. It's obviously dangerous, like all true poetry.

Like most outsider poetry, Olsson's is a flesh-scream with markedly underdeveloped scar formation; it has neither causality nor experience, but is, rather, a continual wound. And where this wound rules, things happen. There's swarming in the flesh ("and how I break myself in the dress," as Olsson writes). In the muck of the wound true action ferments, obscene and barbaric. Olsson goes further than anybody else in this breaking up and drilling into the scar. Things get worse and worse. In *The Sacrifice*, she goes so far that even the most "radical" or maybe neutral literary critics start to get peevish: "Incomprehensible!" Yes, naturally, the results will be incomprehensible when one starts to speak in a new (human) language. What the hell did they expect? That Eva Kristina Olsson would sit still and grow pale on the wall and act civilized? But there's a large charge for breaking. It means making a sacrifice. And now she's gone all in. Now she's lost in there. Where will this all end? In the wound's timeless, worthless semiotic depth. In her next collection will Olsson be able to use this mind-warped comet's journey through an inverted universe to bring meaning to a place where there was no meaning? Will she go farther into the wound? Like Lovecraft in his tunnels, except that here's only the omnipresent death.

Sometimes the Kristevan/Lacanian semiotic mutterings may seem hopeless – the "female no" and the pregnancy as refusals to linguistic lawfulness and signification adjustments blah blah etc. There are many thoughful female writers who work on this *no*, who poke around in the wound. There's nothing bad about that, though they keep to certain limits. But Olsson goes further than that: she is apparently making herself insane. She does more than shout *no* to the scar; she tears it loose from the skin like a gross old Band-aid and lets herself get sucked into the wound.

Beyond the scar lies being mold, being stone-dead. But is it perhaps this wound-scream that will lead us in backwards toward the excluded third way as through a black hole? Will we ever be surprised? When the fuck will something truly new take place? And if so, what will it be?

(1995)

IT'S NOT ACCEPTABLE TO BE A FATSO

The first trap every poet must avoid is narcissism.

You can get pretty far writing in order to be loved (or hated), but it's a betrayal of poetry whose transformational power takes place on a wider front than individual satisfaction. Both writing and publishing poetry require having the strength to be alone, to withstand the sorrow of never being able to reach a symbiotic relationship with one's inner, called-out-to reader—the impossible longing of creation.

To read poetry is the same thing: to be broken up, left abandoned, torn loose from the preconceived ideas one may have gathered over years. Poetry is about letting oneself be damaged, hurt, it's about suffering through the least common version of reality, the version in which it is NEVER possible to recognize oneself again. The version in which all is pain.

The paradox: Intimacy is not possible if one tries to avoid these insights. Communication in poetry is about the path through one's own impossibility.

Literary criticism should not be practiced by anyone who has not gone through psychoanalysis, myself included. One has to be suspicious of one's own hidden motives, though almost no critic is. Literature is defined by and derives from an inner life and it is only through one's inner life that one should encounter literature.

That every literary work of quality communicates both on the conscious and unconscious level is obvious. One has to recognize one's own limits, and now I'm talking about the craft of pure courage: to feel one's own horror and weakness, one's own vanity and dreams. To be a critic is or at least should be a self-revealing activity, and the first question a critic should ask before a text is always personal: how is my unconscious reacting to this, and why?

The second question is: Have I been able to disregard my own preconceived ideas about the author and the subliminal messages she is sending out even with the most discreet marketing? Have I also thought about my own perverse relationship to the author's position or non-position in the social theater? When one has gotten past one's suspect reactions to a certain material, one has all the right in the world to be harsh on the text if one discovers hypocrisy, cowardice, or flattery from the author. As an author one also has to be able to handle being criticized by serious critics.

I hope for poetic articulations that are aggressive, baroque, and esoteric; I prefer ridiculous and embarrassing to perfect. In the literary marketplace, which is dominated by the aesthetic and social ideals of the upper middle class, it's not acceptable to be a fatso in any sense - one adjective too many and you're gone. It's a persistent cliché that the sober, quiet, and elegant text – or, more broadly, the "simple" text— has more to say than the noisy text. The fleshy, screamy, and overcharged, the vulgar, desperate, and pathetic are so taboo in our culture that I can smell a rat. The right to an inner life, to slowness and laziness, and to the time it takes to travel to other worlds are the most important political messages in this time. Work-shyness is the strongest political act. Here poetry is an irreplaceable guide.

The text of the future is inward moving, fractalized, and non-linear. I hope it bites off more than it can swallow.

(1998)

A CISTERN OF SORROW

When contemplating water, it's impossible to start from any other place than the planet Solaris in Stanislaw Lem's great sad novel about eternally-dying life. In *Solaris*, strange memories are brought into the desperate present. The water world is a beautiful assemblage, constantly shifting shape. But for humans, those profoundly water-based creatures, this constant shape-shifting causes pain because there's nothing to cling to. For humans, death is the end of life, a shadow-wall that stands in sharp contrast to their own lives. For humans, there is no greater fear than the fear of a world where it's impossible to die.

On Solaris, the dead return. They're almost the same as when they were alive, yet frighten the very people who generated them through their love. Human consciousness cannot receive such a gift, such a clear proof of the fundamental risk in re-generation: to return, but with a twist, as in the spirit of neither-nor.

In *Pet Sematary* by Stephen King, a small child returns in a similar way. The level of physical horror is higher and the dead monster reaches repulsively wonderful proportions. But who wants to see the horrible Valdemar in Edgar Allan Poe's short story wander around on the streets hypnotized into a zombie state? We dream of being able to always go down into the same river, to be subsumed in the same god – especially in religious thought. But do we really want to experience such a return?

From a human perspective, the ocean on Solaris is a sorrow space, a gigantic magnification of the sorrow space women carry inside us: the uterus. The uterus is a sorrow space in light of contemporary society's treatment of time: birth is a loss in the shape of ceaselessness. The child never stops being in the fetal stage; the pregnancy carves memory traces and memory language into the woman's body. Once pregnant, always pregnant. The uterus-memory is a cistern of sorrow, a collision of two ways of being human and two forms of perception: the social and the asocial.

When I was pregnant with my daughter, I stopped working immediately and stopped seeing most of my friends. I went into the deep-end quite concretely – by daily spending time in water, but also spending time remembering, which poured out of me from a sense of loss I didn't think was possible. I let the sense of loss ravage the knowledge that when one is pregnant dying is not allowed. There's no bottom there. It's a state in which we have no choice. It's impossible to destroy or end this state. Pregnancy is a temporary immortality, the only mental space where a person can survive entering into a continuous feedback loop.

Now I know that people may accuse me of hormone-glorifying essentialist feminism of a vulgar-Jungian type. I disagree: that perspective is uninteresting in this case. The interesting thing to think about is how such strange possibilities for thought emerge from a state which is actually so normal: expecting a child and existing in that extremity, instead of the more functional strategy of trying to normalize it.

The idle, braindead inwardness of pregnancy awakens an almost inhuman sensitivity, with which it's possible to feel thinly distant, almost telepathic movements in the amniotic fluid, beyond the normal butterfly fuzz of floundering and the drone of nudges. This embodied thinking is possible for all (except men) and can perhaps open up new perspectives on the bizarre goings-on in the uterus. For example, the fact that the fetus can see.

The fetus can use its eyes early on, but it's an inward gaze, an inward film looping as the eyes turn toward the brain. What kind of film is this, watched by eyes that see without experience of sensory impressions, this water-film that glows through a flexing transparency? When the pregnant woman closes her eyes, perhaps she too can see deep into her strangely pumping doubleness, a body in which two hearts pound and two brains think. What she can see is the ocean on Solaris.

Water-existence is not just a beautiful realm of dolphin ether-bodies in harmony; it is a world beyond our normal categories, impossible to comport with notions like *humanity* or *beauty*. There is safety in this eternal now, but also a vertiginous abyss of threat: we cannot breathe the air produced where that intransigent flow rolls and runs, since we are both mortal and immortal in the same body: a beloved enemy made almost entirely of water.

It is a water that is more than us, a prehistoric water that has been everywhere, an inward-depth of eye-water that has seen and experienced everything, even that which in our daily surface persists as the incomprehensible, the strange.

(2000)

JOY PACKS AND PARADISE MORONS

Joy is banal and horizontal. It has no depth. It's shallow, irresponsible and full of clichés. It does not want to think. It has no organization, which creates an opening for the imbecilic, without any plans. The goose wants to do its own thing while the water drenches it. The sheep are full and just want to hang.

Joy is often private, and the private is the low-water mark in all socially relevant literature. I have two girls. One is in daycare, the other an infant. The most important thing about them is that they have not become human yet. Humans are aware that time passes. Humans understand cause and effect. The beginning – *peripeteia* – death. Thesis + antithesis = synthesis. Most of all: what can and cannot be combined. One cannot have the cake and eat it too.

My children haven't understood those things yet. They are operating according to completely different explanatory models. The little whatchamacallit is basically wordless except for "Äggu!" and "Häattä!" (Is it Finnish?) and "Hej då!" The big ray of sun "grottofies," "sniskilates" and "elephantasizes." Through her "frace" she looks out on a world of "humanistars." Sometimes anxiety tries to sneak into this world, but it is held back like this:

Then I got some blood.
No no no!
No blood came.
It was just captchup!

Blood is transformed into captchup. I find myself in an insane asylum where suffering is put aside. Everything's loosely connected, like in a game. There is no chronology, no overview. Tomorrow might surprise us by showing up before yesterday. But this lack of structure does not lead to horror, for in the midst of the chaos there's a safety, anchored in nothing: in the order of the small things, in the little fiddlings and prattlings, in the now, in the embrace. This is an idiotic paradise.

This paradise is suppressed, not the least in literature. In the intro to literature classes we quickly learn to disregard paradisiacal writings. Writing joy doesn't work. Dante is of course the first example. Paradise lacks drama, while the depiction of suffering has an obvious arc, a rhythm and structure that represent what it means to be human.

The serious text suffers. And it's male-dominated: Society's and the greater public's great tragic arc is the domain of patriarchy. The private is something women do without making too much noise. These private acts have not qualified as being officially valuable.

But those women (still the majority) that are responsible for the small things – the home, chores, love – are not children. They do not live *in* but *with* paradise. Those who think otherwise are probably modernist sneak-machos who turn women into mystical beings through oppressive idealization.

Those who have to take care of children have to shape up and act grown-up, which adds a complication: in order to get back to joy, a grown-up person has to pass through the hassle, the pain and suffering. The outsider's perspective allows one to see suffering in the paradisiacal. This suffering derives from the relationship to time, the insight that the clueless little idiots are heading for a great loss. Many depictions of childhood get stuck there – they become whiny, as the grown-up feels more sorry for themselves as children than the child did when the child was at one with its childness.

When one tries to write the paradisiacal everyday, there are many possible approaches, all of them with pretty low status in high-brow literature:

- the children's first words, their invented words

- idiomatic sayings, clichés, platitudes
- kitsch
- literalness
- the obvious themes: home, family life, undramatic love, nature

As a grown-up, the ability to take pleasure in the small things of childhood demands both an enormous independence and the capacity to be dependent. It's like practicing changing gears in a car when you first learn how to drive. It's not a polarized situation – joy and suffering are not opposites. Joy is the first stage in the movement toward the suffering of becoming chronologically grown-up. After that, joy is a singular way of being that can't survive in society, where it's exploited and perverted. But when joy takes place on its own terms, it is timeless, opening up toward an excluded third space, a kind of smooth space of creativity, untimespace. The idiotic idyll is a space beyond joy and sorrow. When it is transferred to objects or images one can call it kitsch. An exit where one least expects it: into a hierarchy-less-ness where one utterly doesn't give a damn about money, expensive stuff or self-righteousness.

Paradise wrecks the scale between evil and good and invents the wordless. Thus it makes its own humanity impossible. It rebels against the pretend-important, against self-assertiveness, city SUVs, hurry, men who absolutely have to go around making decisions.

The most common human misunderstanding about joy has to do with permanence – that joy would be possible in normal time. But it doesn't work like that. Joy transforms time. Joy makes the moment wider than an entire human life because it's a moment that stops and sinks into a total presence. For example: the mayfly. With this in mind, it feels important to steal joy back from those who have monopolized it as a fantasy about eternity or a guarantee of safety, for example Christians and banks. Both perpetuate the lie about eternal joy. Open any Christian pamphlet and you'll see lions, tigers and people walking around in blossoming meadows enjoying each other's company. Or open a bank pamphlet and you'll see models in outdoor clothing cuddling with airbrushed children while the sun seeps in through the harmonious flower pots.

Steal back the yellow, happy, pathetic joy – through children or language. Paradise is an undepicted path of rebellion. It uses the pithy un-words of the happy pack:

Letters
with and without
sailfeet
With steady
and not steady
But can stand

(2004)

LANGUAGE AND MADNESS

Language is a living being. I think that language came before humans, not the other way around. The cells of language hovered above earth looking for a host body. It tried to take over dinosaurs and fishes, but that didn't work, they were stupid or the muscles in their speech organs were not developed enough. Then came a human. The invisible, potential words attacked her, like mosquitoes that know that they need blood and have waited for thousands of years for the first human to evolve.

It might not have been a particularly logical language; more likely, it was paradisiacal and timeless, a kind of happy babbling for the sake of babbling, a kind of music. But with time it became more descriptive instead of creative. It became instrumentalized. It began to strive toward developing the future instead of standing with both feet in the joy (or angst) of the fact that the *now* can be expressed and come into existence through words. For some reason, language became both more stupid than it really is and more self-conscious and anticipatory than was good for it. I don't think the whole of humanity was to blame for this development, but rather a very specific segment of humanity, those who benefitted most from the patriarchy—that is to say, men. Since the patriarchy is all about power, those who took control over language made language a means to maintain power. That's how the first war came to be and the first colonies, chronology and narrative structure, the social order we still have today.

But it's not my intention to paint a binary of evil and good here. Perhaps language needed to be tamed in order for us to become more than mayflies. It's too bad language had to be transformed into a market-economy power apparatus for pleasure-opposed morons.

This isn't biological essentialism. There is no natural femininity that inhabits the lovely right half of the brain. But for those who don't want to carry on with patriarchal activities for days on end (usually women, since they don't earn as much for their patriarchal activities as men and have to fight much harder to reach the same goals), for those who don't want to devote themselves to work and money, only one task remains, whether they like it or not: to take care of and communicate with children. Or to take care of their own childishness through playing, puttering and creating things that are not considered important. The playground is the garbage heap of the patriarchy.

What, then, is the relationship between madness and games? Is language a game?

Language contains madness. But one has to be careful not to romanticize psychotic language. One wants to be on the side of madness when one plays with these kinds of thoughts, one wants to be more or less surrealistic. The chronological and rational is boring and uncreative. But at the same time: who wants to be uncontrollably insane? How awesome is that? Sure, we love the mad, but only if we ourselves get to decide the conditions for what the madness will look like.

Children are mad in a fun way, because they are pretty modest in their madness: they play around with their food and scream guttural and enchanting oddities, they have no concept of time and say "hello, ax-handle" without any care. But they do have some sort of control. They do not walk around free on the streets practicing their freaked-out version of reality. Somebody always keeps an eye on them so they don't kill themselves or others. Children are ultimately not good at heart but just physically inferior, which makes it easy to make decisions for them without having to employ four strong men to put them in straightjackets.

But a screaming, real-life-size psycho who scarfs down feces and foams at the mouth and stabs wildly around him with a knife? In that case, society says, *we* don't want to be like *that*. *We* don't want people like *that* around. When people speak out in favor of a life of madness, they mean the cute, nice madness, not the disgusting or dangerous kind. The disgusting and dangerous kind is prioritized in language but not in life.

In language, depressing, dark suffering has higher status than the nutty, pathetic and everyday-life-bizarre craziness. Great men write great tragedies or long descriptive epics out of great universal psychological suffering. Great women write great tragedies or long descriptive epics out of psychological suffering.

Then there are some ridiculous women who write about their ridiculous babies, about their small duties here and there, but there are only a few of those. Books based on great existential madness are a kind of norm. When, in more recent years, there have been some books about motherhood in Sweden, the critics have shoved them into a special little section and said yes, yes we have seen this now, motherhood has become a trend and it's been done. Why? It's a subject matter that is worth just as much space and treatment as the great existential questions and the inferno crises of all kinds of old windbags. Motherhood is one of the most overlooked subjects of 20th century literature: that cute, paradisiacal madness. The mother's relationship to the baby is the root of language, madness, and complexity. None of the great serious works would have seen the light of day without the tracks that were inscribed in the earliest mother-and-child relationships. Life is based on the irrational and noisy language of this crazy little symbiosis.

This symbiosis is a positive psychosis between the mother and child (or father and child in the few instances where the dads dare to take time off to raise the kids). It's a lesson in love. The world is no longer simply a singular, devouring infant, but an infant that grows through interactions. In order for love to develop, there has to be distance. To feel love is only possible if one realizes that symbiosis consists of two people. Love is an automatic split into *you and me*. *You + me = we.* If one is involved in the traditional, patriarchal psychosis there is no *we*. Instead, *I am the world*. In the great, self-righteous male despair there is no *we*, just one bloated *I* that swallows everything that moves.

When interacting with the mother, the child's madness becomes a democratic madness, in difference to the male, ego-trip inferno crisis.

And who is going to depict this small, spaced-out joy? Who has the language to express it?

It can be found in language of the lowest literary status, in the child's first words, in everyday chores, in banalities and in private life, a subject matter that is ridiculed in serious literature.

The paradisiacal is neither good nor evil, but it is an overlooked path of insurgency for language. I think that this joyful, worthless madness is the next big literary and human revolution. And it's free.

(2004)

TSUNAMI FROM SOLARIS

Case #1

Blindwalker[1]

Glitterflakes are filtered
The moon walks on spider legs
with the giant brain's
intestinegray condensation
on fog glass

Lead-worm-broken strokes
drill tear channels
new tearyear's bodystub
The shingle-belt's
soft tube
around midnight's waist
from a thumbed
hour glass

The soil-tunnels' blindwalk
The vertebrae skewer skewers
earthwormpretzel ties
crawlsick eights
pale creature
eyeless
and without tooth

The slagwormdeathquiet dung scream
presses giblet skeleton
Hacked the wreath careens

---

[1] "Blindgångare" is the word for unexploded ordnance, for example mines or ammunition; it literally translates to 'blind walk'

unstickified backslung
around its stabbed half

The bait's suction cup lips
oxygen-poor rumpstump
The umbilical string is biting
The spider crank hulling
Freshen up the loss
Firework the losslust

Eviscerate shame
on dry land
Offscaling, confetti
The nuggetcatch uplunged
The throwup of jellies
fatsludge clumps
sand

Offensive albino
The life spark is embodied
cramps to the core
life-sparking star
on the eye's blindbottom
of a newdead
timeless corehealthy
deepseafish

Case #2

Wave in Pippi Movie Upset Viewers

The animated movie "Pippi in the South Sea" was shown on Swedish Television Channel 1 on Sunday night. The movie included a sequence where Pippi and her friends are washed overboard by a giant wave caused by an earthquake. This lead to many reactions.

"We have received many complaints. I hadn't watched the movie beforehand, but if I had done that I would have excised that part of it," said Bella Stjerne, executive of children's programming at Swedish Television.

"We have returned to normal programming after the catastrophe, but we simply have to be more attentive. At the same time, some children may find consolation in Pippi surviving. These are important tradeoffs for us to consider," says Bella Stjerne.

(*Dagens Nyheter* 1/25/2005)

Case #3

Swedish Radio

Swedish Radio observes three minutes of silence for the victims of the tsunami. Traditionally, a minute of silence consists of just that, a minute of silence. But in this case the silent minute is interrupted by commentators informing us that we are listening to silence:

"Here in Göteborg people are standing with lowered heads in the mall. They are silent..."

"Here in Linköping Cathedral everything has stopped, people are silent..."

"Here at the restaurant Vindas in Umeå, people are silent. Plates of the day's special – foccaccia with onion and tomato – are cooling on the tables..."

*

The poem "Blindwalker" was originally commissioned by *Expressen*'s culture pages: I was asked to write a poem for the new year. I handed the poem in the day before Christmas Eve. When, on the day after Christmas, I heard about the catastrophic tsunami in Asia, I thought about the poem. Earlier I had thought it was pretty honest – a hopeful image of an earthworm skewered on a hook who drowns and becomes bait for a fish, which in its turn dies, is born and transformed on dry land. Now it felt tasteless and mocking, as if I was making fun of the victims.

The mockery would derive in part from the sea imagery – not that it's exceptionally provocative, it isn't. That's my point: almost anything that is at all depressing or ambiguous can become offensive when reality is transformed into trauma. When reality exceeds the poem in misery, ambiguity is no longer welcome.

I realized that the worm-drowning could be associated with the broken bodies of the victims of the tsunami. The fish with its albino-flesh, flushed up on the beach, may invoke the image of a Scandinavian corpse. Sure enough, I received a call from the editors. They used to like the poem but now they were thinking like me. The poem could now give off the wrong associations and be pulled into grief's orbit.

As a result of this, I started to ruminate about poetry. I started to think I was an evil and horrible person for writing disgusting poems, but most of all I felt bad that I was writing poems that were pointless in times of trouble. I was writing poems about the dangers of the world which became useless when that danger actually struck. Were the only poems worthwhile in the long run the nice, warm poems of consolation, crisis management, community? Should one in certain moments avoid writing about or depicting angst about mortality? I'm not being ironic when I ask: Is it Lasse Berghagen[2] we need when reality really hits us?

[2] Swedish entertainer

In the days between Christmas and New Year's, Swedish Television worked overtime to produce – or if I am being generous, document – a collective sense of national grief, a mixture of true compassion, existential angst and a perverse kind of regression, in which the king, a lady from a travel agency with teddy-bear-brown eyes, and a gang of priests replaced the mother's breast. Images circulated wildly; TV programs were pre-empted to make space for a loop of images of the wave caught by amateur photographers and also, gradually, images of corpses – "bodies," not the "dead" (media colluding with the priest lobby – the souls were probably presumed to be with God)—that lay rotting in piles everywhere. There was a fine line between what was allowed and not allowed to be shown on TV: authentic reality (what can be captured on film) was OK, but nothing that did anything else with the imagery, because that was at risk of generating "wrong associations." This prohibition was not limited to depressing or bizarre iterations, making any kind of version disturbed. Still a month after the tsunami, fictional images were so much more threatening than documentary images that it was wrong to show an animated, invented wave in a children's program about Pippi Longstocking without first engaging in "difficult considerations."

Or, to return to the example of the chattering minute of silence on the radio: One might defend oneself by laughing at the absurdity of it, but ultimately I have to confess that it all makes me feel ill at ease in a way I can't quite understand. It is a fictive copy of the silence, a reproduction that chafes and fucks up. It is not authentic. It has margins where worrying things happen.

The tsunami is itself a kind of flipped-out copy: a UFO wave that has nestled its way into the normal human ocean. It is a wave that doesn't understand how to be a normal wave and that throws itself against tourist beaches and the un-mundane people's plagiarized, artificial paradises (Normal inhabitants of these countries worked, were not living in artificial paradises. So in their cases, perhaps death was closer to reality). It's as if the tsunami wave was set in motion by the ocean on the planet Solaris in Stanislaw Lem's novel.

In *Solaris*, the copies of the dead loved-ones are themselves anxious about being erroneous associations: it's as if the tsunami had feelings and was conscious of its own terrifying task. In *Solaris*, the memories of the dead are materialized in a dark collaboration with the unfathomable, threatening, and probably cognizant ocean: figures of pain and love return in physical form, almost-but-not-quite human. "The bodies" are alive but only partly in contact with the living souls.

In the poetry collection *Solaris Korrigert (Solaris Revised)* by Øyvind Rimbereid, the expanded mirror world of Solaris has moved down in the deep sea somewhere off the coast of Norway, and it's described in a plain and comprehensible but deformed language. The I becomes a copy of itself, a copy without a body:

> IT "your duplicat Ei" shal be, say they
> so humans aftr me can no who I was.
> BUT wat wuud it self denk on?
> WAT wuud it denk on me
> who once original existens was?
> WUUD it on me denk wit delyte
> or wit sorro?

But the ocean crashes in on the way down and the poem's *I* never reaches its destination. Or is it the other way around? It's an open question. But how does Rimbereid have the patience to revise Solaris when Lem's novel is already an edited version of the human, a version one can barely stand? When Lem's copies, for all their deformations, tell us more about real feelings than any supposedly neutral documentary?

Solaris is poetry as matter. The tsunami is poetry in its essence. How can one carry further the cruel poetry-as-matter when safety collapses and that which is a version of reality is more horrifying than the real? Why is similarity scarier than authenticity? Why is the copy more dangerous than the original? Why is the poem such an insult to the cruelty of life itself?

(2005)

LANDSCAPING

In my brain there's something resembling Miman in Harry Martinsson's sci-fi epic *Aniara*: a machine projecting films, holograms, or maybe instructional dance videos. I don't know how I became hooked up to these worlds, but I receive them with fascination mixed with horror.

The landscapes are always first to get my writing going. The content of my poems is driven by the landscape's atmosphere, color, smell, light, and, not least, weather, temperature, and the four elements. Is it an icy cold, airy, high-blue glacier atmosphere? Or an underwater world of deepblue breathing sounds? I don't understand how it works, I maybe have a highly-functioning tendency for hallucination, but when I write it's like I sleepwalk and suddenly find myself in that place.

What do I do there? I don't know if my poems answer that question but they try to at least understand what the space wants.

But I don't mean that these landscapes are there as some kind of jolly road sign existing solely for my narcissistic confirmation or for the sake of safety. The place is not a sign, not a symbolic sphere. The inner journey has nothing to do with religion or the New Age, it doesn't satisfy any need for meaning. There is no God who serves up riddles to help me find the meaning of life. That's not how I look at my poetry.

My ur-landscapes do not want anything from me; they are in that sense cold worlds. They exist, but not for me to exploit. By good fortune I can sometimes catch glimpses of a reality without voracious, corrupting people. Yet I cannot label such a landscape as some kind of nature reserve, or worse, a zoo.

The inner landscapes ignore humans. They do not communicate like us. In this way, they are reminiscent of any place in nature.

My landscapes are not static. Through my books I trace their evolution. My landscape journey began in a real but also decidedly unreal place, in a godforsaken part of Norway. Every summer during my childhood, I traveled from Tensta in Stockholm to a place by the coast outside of Trondheim. We had (and still have) a mountain hut without electricity or water near the village where my dad was born.

The contrast between everyday life in the suburb and on the mountain was of course enormous. On the mountain there was no civilization, no telephones. The marsh outside was suctioning and bottomless, the mountains rose all around us, magically huge and prehistoric. The contorted pines and mountain birches were alive in every little twist of fiber-muscle.

But this ur-place was not just beautiful, it was also full of ghosts. One time my sister and I heard a horrifying inhuman scream in the middle of the woods. That night I saw a face pressed up against the kitchen window, but there was nobody there. When I was three, according to the family myth, I was found sitting on a bridge speaking fluently in the local dialect with my dead grandfather. Not that I believe that, not for real. But, at that age, "for real" is a far subordinate parameter in world-building: ask a little child what is a dream and what is reality and they will, thankfully, have zero clue. A child's reality is built on fantasy.

As time passed and my personal sense of reason developed, things happened to both me and the place: modernity was pushing in. The old farmers in my family died out. Rich Norwegians, driven by a perverse desire to burn their oil money, began drilling into the mountains and massacred a big part of the endlessly beautiful road that lead to our magical place. There were no longer winding roads with only crumbling rails as protection against the chasms where the ocean crashed back and forth in a forceful sweep of ebb and flood. Instead we got ugly high-tech tunnels with hissing ventilators. And then some damned morons built houses with solar panels out on the islands in the desolate marsh. Some gang of idiots decided to climb along the 400-meter-high mountain façade behind the house. Were they bird-watchers? Parachutists? And right next to our land, at the end of a half-mile-long part of the mountain with the prehistoric name Ulvstubakken, they suddenly built a rest stop for tourists. Ulvstubakken with its prehistoric name: perhaps my ancient ancestors' skeletons are rolling in their graves there, beneath layers of German toilet paper.

The inner Ur-Norway has been pushed aside, devoured by adult reality.

It was maybe just as well. Other worlds appeared, other inner landscapes loosely connected to the outer world. I wrote a book called *Uppland*, where the action is based on the almost ridiculously happy and idyllic summer nature of Middle Sweden. And in the book after that, *Loss*, the landscapes began to flow through the TV: the terrible wave of the tsunami, the deepsea where the corpses of small children who had been torn from their parents' arms and sunk down to the aquarium-calm sandy bottom of the ocean.

The poetic landscapes slowly take shape and fall into place, like in a dream sequence, but in my case the place comes closer and closer to reality.

I'm starting to almost come close to the situation where the landscape of the poem is the same as the landscape where I live: the farm Petters in Övernjöte, Njutånger, Hälsingland.

This new discovery will probably shape my next book: The souls of animals, children, buildings and cars are so much more fascinating than vague creations of the imagination. Aliens, ghosts or psychic phantasms cannot exceed the strangeness of a normal, honest farm hen.

In the Hälsingland countryside, the Norwegian mountain austerity and the supercute Middle-Swedish joyscape have gotten hooked into each other. And the world of TV images is not far off, either: I can't claim that it's merely idyllic to follow in close-up the harrowing events caused by climate change and the weather extremes from year to year, something that can barely be noticed in the weatherless fake environments of Swedish cities.

My feet are on the ground now. The inside looks like the outside. I don't know if that's good or bad, but for the first time in my life I don't feel like an alienated alien. More like a crass, practical, cross-eyed hen. The Superreality is here. It's finally real.

(2008)

ACKNOWLEDGEMENTS

"The Wounded Path Toward An Interesting Life" ("Sårvägen mot det intressanta livet") was first published in *Stora Saltet*, 1995.

"It's Not Acceptable to be a Fatty" ("Det är inte accepterat att vara fetto") was first published in *90tal*, 1998.

"A Cistern of Sorrow" ("En cistern av sorg") was first published in *Göteborgs-Posten* in 2001.

"Joy Packs and Paradise Morons" ("Gladpack och paradispuckon") was first published in *Subaltern* in 2004.

"Language and Madness" ("Språk och vansinne") was first published in the anthology *Swinging with Neighbors* in 2004.

"Tsunami from Solaris" ("Tsunami från Solaris") was first published in *Lyrikvännen* in 2005.

"Landscaping" ("Landskapande") was first published in *Provins* in 2008.

The excerpt from Øyvind Rimbereid's *Solaris korrigert* in "Tsunami from Solaris" was translated by David M. Smith.

CONTRIBUTORS

Aase Berg is the author of seven books of poetry, two books of criticism, a YA novel and, most recently, the novel *Haggan*. In 2011, she was awarded *Dagens Nyheter*'s Lagercrantzen for her criticism. In 2012 she was awarded *Aftonbladet*'s literary award for her poetry. Her work has been translated into numerous languages. Five books of her poetry have been translated into English, including *With Deer* and, most recently, *Hackers.*

Johannes Göransson is the author of six books of poetry, *Transgressive Circulation: Essays On Translation*, and the diaries *POETRY AGAINST ALL*. He is also the translator of several books of poetry by Aase Berg, Johan Jönson, Kim Yideum, Henry Parland and Ann Jäderlund. With Joyelle McSweeney, he edits Action Books. He teaches at the University of Notre Dame.

Joyelle McSweeney is the author of eight books of poetry, prose, drama, and criticism: *The Red Bird, The Commandrine,* and *Flet,* from Fence*; Nylund, the Sarcogropher* and *Salamandrine, 8 Gothics* from Tarpaulin Sky; the verse play *Dead Youth, or, the Leaks,* which inaugurated the Leslie Scalapino Prize from Litmus Press; the critical volume *The Necropastoral: Poetry, Media, Occults* from the University of Michigan Poets on Poetry Series; and *Toxicon*, a volume of poems forthcoming from Nightboat Books in 2020. With Johannes Göransson, she cofounded Action Books and teaches at the University of Notre Dame.